THE “WESTER

Part 2: Post-War Years

ROBERT GRIEVES

TURNTABLE PUBLICATIONS

SHEFFIELD

OTHER SCOTTISH TITLES:—

Edinburgh Corporation Bus Album
Edinburgh Tramways Album
Scottish Buses Before 1928
Scottish Electric Tramways

ISBN 0 902844 30 X

Crown Press (Keighley) Limited, Chapel Lane, Keighley, West Yorkshire

INTRODUCTION

At the end of the Second World War, the first operator to be acquired by Western was W. & R. Dunlop Ltd., of Greenock. However, this firm continued working as a separate operating subsidiary, and was not actually absorbed by Western until 1949.

This was a policy decision taken by the British Transport Commission, which had acquired the whole S.M.T. group by this time. Also affected by this decision to wind up the operations of subsidiary companies in 1949, were Greenock Motor Services and Rothesay Tramways. These two companies thus lost their separate identities (the latter no longer operated trams by this date).

The services and vehicles of the Caledonian Omnibus Co. Ltd., of Dumfries, also passed to Western in 1949, having been purchased by the B.T.C. as part of the Tilling Group. The "Caley" had been formed in 1927 by the British Automobile Traction Company to acquire smaller operators in the Carlisle, Dumfries and Kirkcudbright areas in particular. The following year, it became Tilling and B.A.T. Co. Ltd-owned, and then part of Tilling Motor Services in 1942. Western took over operation of all the services apart from some to Edinburgh, which were handed over to Scottish Omnibuses Ltd.

Another important firm, whose buses and routes passed to Western in 1951, was Young's Bus Service Ltd., of Paisley, and its associate, the Paisley & District Omnibus Co. Ltd. "Y.B.S." and "P.&D.", as they were known locally, were nationalised in 1950 upon acquisition by the B.T.C. "Y.B.S." operated its yellow fleet on the longer services from Glasgow, through Paisley and Johnstone to the Ayrshire coastal resorts of Largs and West Kilbride, while the "P.&D." ran the local routes in Paisley.

Also in the Paisley area, "Western" took over operation, in 1955, of the local bus services operated by Glasgow Corporation Transport to Abbotsinch, Kilbarchan, Barrhead and Renfrew Ferry. No vehicles were transferred, however.

In 1965, Rothesay Motor Services on the island of Bute (an associate company of McGill's Bus Service of Barrhead) was acquired. Also in the mid 1960s, the following independent operators were purchased: Clark of Glencaple, Dumfries; Campbell of Gatehouse-of-Fleet; Murray of Stranraer and in 1968 Smith of Barrhead (owned by the Scottish Co-operative Society). In all cases, only the operators' routes were taken over, and none of their vehicles entered service with Western.

The most recent take-over was in 1970, when the bus services of David MacBrayne Ltd. were divided between Highland Omnibuses Ltd. and Western S.M.T. "Western" inherited some of the more southerly services, the main one being the 99-mile route from Glasgow to Tarbert, on Loch Fyne. Four of MacBrayne's small seating capacity Bedfords also passed to Western, and were numbered 1-4 in the fleet.

Over the years, the long-distance services have increased considerably, and in conjunction with the operators in the areas concerned, it is now possible to travel from Glasgow to Dunstable, Ipswich, Bradford, Southport, Paignton and Coventry. Regular summer services to Eire and Northern Ireland are also operated, in conjunction with Ulsterbus and the Londonderry & Lough Swilly Railway Co. buses, to Bundoran, Letterkenny and Londonderry, utilising the Stranraer/Larne Ferry. Services to London, Blackpool, Liverpool and Manchester have, of course, been maintained for many years, and are "Western's" oldest established runs to England.

A word now on "Western's" fleet-numbering system. These have been used on various occasions, but the scheme in use today, which has lasted longer than any other, originated in 1949. This is based, more or less, on the age of the vehicle, and started from 1 upwards. A few numbers have been re-used over the years. Those vehicles acquired from Young and Paisley & District simply had 2000 added to their existing numbers. Since all these buses have now been withdrawn, the numbers from 2000 upwards have been re-used.

As will be seen from the illustrations, the fleet numbers are preceded by two letters. The first of these denotes the depot at which the bus is based, and the second is the company code for the type of vehicle.

Another code now in use on "Western" buses is one which gives the year in which the vehicle was new. The letter A signifies 1950; B — 1960; C — 1970. Therefore B6 would mean that the bus was delivered in 1966, and C3 1973, for example.

The fleet livery is basically red and cream, in which colours all the double deckers are painted, but the coach fleet retains the smart black and white which has been in use since the 1930s. The WESTERN fleetname, in underlined capital letters, has remained entirely unchanged since the formation of the company in 1932.

The current fleet strength of the Western S.M.T. Co. stands at approximately 980 vehicles, being the largest within the S.M.T. group.

The photographs which follow will help to tell the story better, and bring it up to date from 1913, where we began in Part I.

Thanks are due to Charles Billette, Gavin Booth, Robert Connor, John Fulton, Ian Maclean and Hector Souter, who helped to provide the illustrations.

2206 (NAG 112G), the first of "Western's" 1969 batch of Bristol REMH6G coaches with Alexander bodywork for the Glasgow-London service. These vehicles are all finished in the black and white coach livery.

Taken over with the fleet of Dunlop, Greenock, in 1949, was CK 3959, a Leyland Tiger TS1 which had originally been in the Ribble Fleet. It became 630 with Western.

Dunlop's TR 6216, a Leyland Titan TD1 of 1929, was originally with Hants & Dorset. The body was built by Croft of Glasgow just after the war. Its Western number was 637.

BAG 117 was an A.E.C. Regal with Brush body, new to "Western" in 1947. It was transferred to the Dunlop fleet, and numbered 623 on its return to "Western" in 1949.

XS 3490, a Northern Counties-bodied Albion Venturer of 1934, was 88 with Young of Paisley, becoming 2088 on acquisition by "Western". It is seen in Clyde Street, Glasgow.

Seen in the livery of Rothesay Tramways, SJ 1049 was a Burlingham-bodied A.E.C. Regal new in 1946. It became number 770 when the company was absorbed by "Western".

XS 3328, a similar vehicle to that in the previous photo, is shown here as 2086 in the "Western" fleet, working on the Largs local service.

A Leyland Titan TD4 of 1936, with Northern Counties body. XS 4104 was Young's Bus Service number 100, seen here as "Western" 2100, at Port Bannatyne on the Island of Bute.

Y.B.S. number 110 was this 1938 Albion Venturer bodied by English Electric. After take-over by "Western" it received a new E.C.W. body.

"Western" 2157, one of the Guys in the next photo, seen here leaving the ex-Y.B.S. Johnstone garage, which has since been replaced.

Part of a fleet of Guy Arabs delivered to Y.B.S. in 1946, with Northern Counties bodies. They are seen here inside the Young's Johnstone garage.

The only exposed-radiator Leyland PD2s in the "Western" fleet were those acquired from Y.B.S. in 1951. 2181 (XS 6908) built in 1950, is seen in Paisley in 1963.

Ex Paisley & District number 297, a 1949 Guy Arab with Guy bodywork, seen on a Paisley local route at Foxbar.

Also ex-Young's, 2200 (XS 6983) was a Maudslay Marathon of 1950 with coachwork by Scottish Aviation of Prestwick. It is seen here in Rothesay on an island tour.

Delivered to Young's in 1950, this Daimler CVG6 was bodied locally by Brockhouse of Clydebank. It became 2202 when taken over by "Western".

Another Daimler CVG6, 2210 (XS 6753) was also new to Y.B.S. in 1950. It, however, was bodied by Northern Counties.

BAG 88 (290) was an A.E.C. Regal of 1946 with Burlingham body, seen here leaving Glasgow for Gourock.

The same bus as that in the previous photo, after having been rebodied by Bristol in 1954. It is seen in Castle Douglas, bound for Kirkcudbright.

Part of a fleet of 17 Northern Counties-bodied A.E.C. Regents delivered in 1946. They were photographed at Inchinnan Garage, from where they operated for most of their lives.

Delivered new to "Western" in 1946, 268 (BAG 385) was a Northern Counties-bodied Guy Arab. It is seen leaving Ayr bus station for Catrine.

This was a Leyland Titan TD4 of 1937, bought from Sheffield Corporation in 1948. It received this new Burlingham coach body before entering service with "Western", who numbered it 569.

379 (BCS 438), a Daimler CVA6 with Northern Counties body, new in 1948. It is seen in Ayr.

487 (BCS 337), a Leyland Titan PD1 with Leyland body delivered in 1948, seen at the Clyde Street stance in Glasgow.

VS 4367 of Greenock Motor Services, which became 728 when "Western" took over. This Leyland PD1 of 1948 is seen with its original Strachan body. It was replaced by a new E.C.W. body in 1952.

Leyland PD1s 507 and 513, of 1949 and 1950 respectively. When new, these had Strachan bodies (see previous photo), but as early as 1953 these were replaced by the E.C.W. bodies seen here.

604 was a Leyland Tiger PS1 of 1949 with Alexander body, seen here on a Greenock local service.

VS 4433, a Guy Arab of 1947 with Guy body, seen here in Greenock Motor Services livery; it became "Western" 652.

872, an Albion Venturer of 1935 with Croft body of 1946. This was taken over with the Caledonian fleet, but was originally with Glasgow Corporation. It is seen at Lincluden on a Dumfries local service.

823, an E.C.W.-bodied Bristol L5G of 1947, acquired with the Caledonian Omnibus Co. of Dumfries in 1949.

Seen in Dumfries, 846 (DSM 453), an ex-Caledonian Dennis Lancet of 1938 with E.C.W. body.

Another ex-Caledonian Dennis Lancet of 1938 was 849, seen in Glasgow. This had a Brush body.

895 (HD 3700) and 869 (GE 2488), two of the oldest buses to become part of the "Western" fleet in post-war years. Both were 1928 Leyland TD1s, ex-Caledonian. 895 was originally Yorkshire Woollen District and 869 Glasgow Corporation.

One of the newest Caledonian vehicles which passed to "Western" is seen in the centre of this group of Bristols in Dumfries. 877 was a K6B, new in 1948; the others are FLF6G 1712 of 1962 and LD6G 1635 of 1961.

11 (BSD 281), an Albion Valiant CX 39N with Alexander body, delivered in 1951, leaving Ayr for Butlin's Camp.

CAG 797 was not given a fleet number. It was one of 16 Daimlers with Burlingham coach bodies delivered in 1948, but which were withdrawn and sold as early as 1949.

A 1939 Leyland Tiger TS8 (175, AAG 114) which had its original Alexander body replaced by this one, also by Alexander, in 1951. It is seen at Dalbeattie.

928 (HWS 935), an A.E.C. Regal IV of 1951 with Alexander body. The coaches of this type were intended originally for Scottish Omnibuses, but were diverted to "Western" and used on the London service.

An Alexander-bodied Daimler CVG6 of 1952 (941) seen at Paisley Cross.

973, an Alexander-bodied Guy Arab UF of 1952, seen on tour at Arrochar.

New to London Transport in 1943, 1009 (GLL 575) was one of a large fleet of wartime Guys sold to "Western" in 1952, when they were rebodied. This one received an Alexander body.

1049 (GLF 692), another ex-London Guy of 1943. This one received the Croft body from a rebodied Leyland TD2 from the Caledonian fleet. It is seen at Carlisle.

The first of a batch of 40 Guy Arab LUFs with Alexander bodywork delivered over 1954 and 1955. 1086 (FSD 464) is seen near Gatehouse on the Stranraer to Dumfries route.

An example of the first batch of Leyland PD2/20s delivered in 1955. This is 1142 (GCS 228) with Northern Counties body.

A Guy Arab IV of 1955 with Northern Counties body. 1178 is seen on a Paisley local route.

1227 (HSD 15), the first of a batch of 20 Bristol/E.C.W. Lodekkas delivered in 1956, seen at Foxbar.

1247 (JCS4), a Guy Arab IV of 1956 with Alexander bodywork fitted with platform doors. It is seen leaving Girvan.

An Alexander-bodied Bristol LS6G of 1957. 1278 is seen near Renfrew.

Seen nearing the end of its express journey from Glasgow to Blackpool, 1385 was one of only 3 Albion Aberdonians in the fleet. It was new in 1957, with Alexander body.

An Alexander-bodied Leyland PD3/3 of 1959 (1524) seen at Abbotsinch.

1608 (OCS 730), a Leyland Leopard of 1960. These coaches had Alexander coachwork with seating for only 30, since they were used mainly on the London service. Two of the older Guy coaches may be seen behind.

The last rear entrance Bristol Lodekka to enter service with "Western". 1647 (RAG 413) of 1961 is seen here at Portpatrick in Wigtownshire.

1666 (RCS 364) of 1961 leads 3 similar Leyland PD3s with Alexander bodywork, picking up dockers in Glasgow.

A Northern Counties-bodied Albion Lowlander (1752) delivered in 1963, seen near Prestwick.

In comparison with the Lowlander in the previous photo, 1758 of the same batch had an Alexander body.

An Alexander-bodied Leyland Leopard of 1964, leaving Kirkcudbright for Borgue. 1894 (AAG 96B).

1950 (AAG 118B) is representative of the last batch of Bristol Lodekkas delivered new to "Western". It is an E.C.W.-bodied FLF6G of 1964, seen near Glasgow Airport.

Seen in Clyde Street, Glasgow, shortly after delivery in 1965, is 1977, one of the first Daimler "Fleetlines" in the "Western" fleet. Alexander 75-seat bodywork is fitted.

2102 is one of the longer "Fleetlines" delivered in 1966. These examples have 83-seat Northern Counties bodywork, and all operate from Paisley depot.

Seen here on the island of Islay, still in its original livery, is MacBrayne number 54 (YYS 174), a Bedford of 1960 with Duple body, incorporating mail compartment. It became "Western" number 1 when taken over in 1970.

An Alexander-bodied Leyland Leopard of 1973, typical of the modern single decker service fleet. 2401 (YSD 361L) is seen in Paisley.

Second-hand Bristol/E.C.W. Lodekkas from several English companies arrived in 1973 in exchange for Bristol VRTs. These two examples, headed by 2434 (PBL 55F) came from the Thames Valley and Aldershot Omnibus Co., Reading.